By Neil McGeehan

Illustrated by Ignacio G.

ISBN 978-1-0878-5428-1

INTRODUCTION

One day, while in High School, my best friend called me and
read a riddle to me; he wanted me to answer it.
It's in this book, called A Ship and Its Boiler. The answer in this book is
the answer I came up with that day.

When we were in college I had a logic course (one of my favorite) and
the text book, by Copi, had an even number of riddles in the front of the book.
The answers to the odd-numbered riddles were in the back of the book.
My best friend was at my house, we were drinking my Dad's beer and
solving all of the riddles. When we got to the last riddle, I answered it instantly.
I still remember what my thoughts were. The next day I asked the professor if I
had the correct answer to the last riddle. I told him the answer and he said,
"You don't have to take the rest of the course." I did anyway and I loved it.

For decades I've been publishing my monthly riddle. It started as a program
on a computer and soon jumped over to email. Every month I email last month's
answer and this month's riddle. I send it to my family and friends.

The riddles I love are logic riddles. Let me give you an example.
You have two coins that equal 30 cents and one is not a nickel.
What are the two coins? You heard, "No nickels" but that is not what was said.
One is not a nickel. Is a quarter a nickel? No. So the two coins are a quarter
and a nickel; a quarter is not a nickel. This is a logic riddle, strict logic, no tricks,
not a game. There's a riddle in this book about the King and the Bags of Gold.
Everyone loves the answer, even if they didn't get it. You will too, it's so logical.
The only time I don't publish logical riddles, strictly speaking, is October, then
I use indirection riddles. They are still logical but I'm purposely misdirecting you.
Here's an example. How many of each animals did Moses take onto the ark?
Zero, Noah took animals onto the ark, not Moses; that's where the logic is.
I included one of these riddles in the book, it's called the Misdirection Riddle.

I need your help. I want the 2nd book to be from all over the United States.
Send me your favorite riddles, to Knowlidence@gmail.com and tell me if you
want your name printed under the riddle, like,
"This comes to us from Pat who lives in XYZ." Tell me if you do want your name
printed, do you want just your first name or the whole name and give me your
City and State. I want the 3rd book to be from all over the world.
The 4th book will be original riddles, written by me.

I hope you love this book. I hope you send me riddles for the 2nd and 3rd books.
Grab your favorite drink and let the fun begin.

Socks

A drawer is full of socks.
In this drawer there are 25 blue socks
and 25 pink polka dot socks.
The socks are loose, they are not in pairs.
The room is completely dark.
You cannot see the color of the socks.
Reach inside the drawer and start pulling out socks.
What is the minimum number of socks
you must pull from the drawer in order to
guarantee that you are holding at least
one matching pair?

A Fork in the Road

*You are a visitor in a mythical town.
Only two types of people live in this town,
Liars and Truth-Tellers.
Liars always lie and truth-tellers always tell the truth.
You do not live in this town;
you are neither a liar nor a truth-teller.
You are on your way to the capital.
You come to a fork in the road.
In the middle of the fork is a tree
and a native of the town is leaning against that tree.
You don't know if he is a liar or a truth-teller.
You can ask him one question,
such that he'll reply
Yes or No.
What one question can you ask him,
to guarantee that you will take
the correct road to the capital on your first attempt?*

The King and Bags of Gold

The king has ten subjects.
Once a month, each subject brings
the king a bag of gold.
In each bag, there are ten bars of gold.
Each bar weighs one pound.

Recap,
the king receives 100 bars of gold each month,
split into ten bags.
The king receives 100 pounds of gold each month.

The jester tells the king that one of the subjects
is shaving an ounce off of each bar in his bag.
The jester doesn't know which subject is guilty.

The king orders all of the subjects to bring him
their bags of gold.
Each subject stands before the king,
in a row, with his open bag of gold at his feet.

The king has a digital scale;
an item is placed on it
and the scale displays the weight.
The king can make only one weighing,
such that he can receive only one weight display.
He cannot put something on the scale,
weigh it and remove or
add something and weigh again;
that would be two weighings.

Using the scale only once,
how can the king guarantee
he knows the guilty subject?

Prisoners and 5 Hats

The warden offered three prisoners a challenge.
The warden showed the prisoners five hats;
two white and three black.
The warden disposed of two of the hats but the
prisoners didn't know which two were disposed.

Each prisoner was blindfolded and the warden
placed a hat on each of their heads.
The blindfolds were removed.
The prisoners were allowed to look at the
hats on their fellow prisoner's but
they were not permitted to look at their own hat.

Any prisoner who could correctly name the color
of the hat on his own head would be released from
prison immediately.
If they guessed incorrectly, their original term
would be doubled.
They could give only one answer.

The first prisoner looked at the hats on prisoner #2
and prisoner #3 and said, "I don't know."

The second prisoner looked at the hats on #1 and #3
and said, "I heard what #1 said and I don't know."

The third prisoner said,
"I am blind. I can't see anything but I heard what
my friends said and I know the color of the hat
that is on my head."
And then he named it correctly.

What color was the hat on number three's head
and how did he know?

Boxes of Fruit

Three boxes are labeled
"Apples," "Oranges,"
and "Apples and Oranges."
Each label is incorrect.
You have to rearrange the labels so they are correct.
You can only look at one piece of fruit,
from only one box.
Tell me which box
and I'll hand you a piece of fruit from it.
Now move the labels so they are accurate.

A Ship and Its Boiler

A ship is twice as old as its boiler was
when the ship was as old as the boiler is.
The sum of their ages is 49.
How old is the ship?
How old is the boiler?

Four Pieces
of Cardboard

You are given four pieces of cardboard.
You are told that each one is
either red or green on one side,
and that each one has either a circle
or a square on the other side.
They appear on the table as follows:

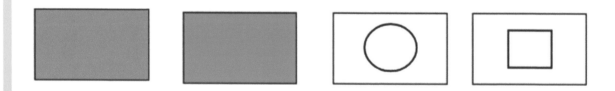

Which ones must you pick up and
turn over in order to have enough information
to answer the question:
Does every red one have a square on its other side?

Diamonds and Rubies

A wicked Queen lived in a palace.
The Queen was so wealthy that her palace was
surrounded by a garden that was littered with
diamonds and rubies.
The wicked Queen's beautiful daughter wanted to
marry the handsome Prince.
The Queen wanted the Prince for herself.
So the Queen took her daughter to the garden.
She explained that she would place
a diamond and a ruby into a box.
Then she would ask the Princess to pick
one of the stones from the box, without looking.
If the Princess pulled out a diamond,
then the Princess would marry the Prince.
If the Princess pulled out a ruby then the
Queen would marry the Prince.
The Queen reached down and quickly placed
two rubies into the box.
The Princess saw everything.
The Queen asked her daughter to pick.
How can the Princess marry the Prince?
She has to pick a stone.

Monkey Business

A rope over the top of a fence
has the same length on each side.
It weighs one third of a pound per foot.
On one end hangs a monkey holding a banana,
and on the other end a weight equal
to the weight of the monkey.
The banana weighs two ounces per inch.
The rope is as long (in feet) as the age
of the monkey (in years),
and the weight of the monkey (in ounces) is the same
as the age of the monkey's mother.
The combined ages of the monkey
and its mother are thirty years.
One half the weight of the monkey,
plus the weight of the banana,
is one fourth as much as the weight of the weight
and the weight of the rope.
The monkey's mother is half as old as the
monkey will be when it is three times as old as its
mother was when she was half as old as
the monkey will be when it is as old as its mother
will be when she is four times as old as the
monkey was when it was twice as old as its mother
was when she was one third as old as the monkey
was when it was as old as its mother was when
she was three times as old as the monkey was
when it was one fourth as old as it is now.
How long is the banana?

Misdirection Riddle

Three men check into a hotel
to prepare for Halloween.
The manager tells them the cost of the room is $30
and each man pays $10.
The manager asks the bellhop
to show the men to their room.
When the bellhop returns the manager informs him
that the room actually costs $25.
The manager gives the bellhop five one dollar bills
and asks the bellhop to return the money to men.
On the way to the men's hotel room,
the bellhop puts $2 in his wallet.
When he arrives at the room,
he tells the men that the bill should have been $27
and he gives the men $1 each.
Each man originally paid $10
and each man received $1;
therefore each man paid $9. $9 x 3 equals $27
plus the $2 in the bellhop's wallet equals $29.
What happened to the missing dollar?

Socks Answer

Pull the first sock out; is one a pair? No.

Pull the 2nd sock out; are you guaranteed to be holding a matching pair? No.

Pull the 3rd sock out. Wait a second…you just figured it out. Good job.

*The third sock has to match either one
or both of the first two socks, therefore three
is the minimum number of socks you must pull
from the drawer in order to guarantee that you are
holding at least one matching pair.*

A Fork in the Road Answer

*This is the riddle I was talking about in the intro, drinking my Dad's beer and
the professor said I didn't have to take the rest of the course.
My very first thought was,
"I can't ask a question about the outside world, like,
'Is the sky blue' or 'Is the grass green.'
Therefore I have to ask something from inside,
I have to ask a question about the answer to the question.
I then I got it.*

*Point to one of the roads and ask,
"If I were to ask you if this was the road to the capital,
would you say 'Yes'?"*

*Try it out. Let's say there are two roads, A and B and
A is the correct road to the capital. Point to A and ask,
"If I were to ask you if this was the road to the capital,
would you say 'Yes'?"*

*A truth-teller would say "Yes". Ask a liar that question
and they truthfully would lie and say "No."
But they can't tell the truth, so they lie and answer, "Yes."*

*The point is, no matter which road you point to,
they both give the same Yes or No answer.*

The King and Bags of Gold Answer

When I first heard this riddle I thought,
"If I take a uniform number of bars, like two, from each bag,
that won't tell me who is shaving an ounce off of each bar.
Therefore, I have to take different amounts from each bag."

The king takes one bar from the first bag, two bars from
the second bag, three bars from the third bag … nine bars from
the ninth bag and ten bars from the tenth bag.
That's 1+2+3+4+5+6+7+8+9+10 = 55.
The king placed all 55 bars on the scale.
If all bars weighed one pound, the weight would equal 55 pounds
but one person is shaving an ounce off of every bar in his bag.
So every bar in the thief's bag weighs 15 ounces.

However many ounces short the weight on the scale is from
55 pounds, that's the number of the person who's bars are
an ounce short. If the weight of all 55 bars is one ounce
short of 55 pounds, then the bad guy is the first subject.
If the weight of all 55 bars is two ounces short of 55 pounds,
then the second guy is shaving an ounce off of each bar in his bag.
If the weight of all 55 bars is seven ounces short of 55 pounds,
then the seventh guy is the crook, etc.

Admit it, you love the answer!

Prisoners and 5 Hats Answer

There were three red hats and TWO WHITE.
Two hats were thrown out. If anybody sees TWO WHITE hats,
then he KNOWS there is a red hat on his head.
Since there were only TWO WHITE hats to begin with,
seeing both of them assures the viewer that there are no
more white hats left; therefore he has a red hat.
Right away, we know that NOBODY SAW TWO WHITE HATS,
because number 1 and number 2 said they didn't know what
they had; therefore, neither of them saw TWO WHITE hats.

Let's say number three has a white hat. Who will see it?
Both number one and number two will see it.

If #2 sees a white hat on #3, then

#2 would KNOW that his (#2's) hat was red, because,
if he's looking at white on #3, then if his (#2's) hat was white,
then #1 would have seen two white hats (one on #2 and one on #3)
and we already know that NOBODY SAW TWO WHITE HATS.

Therefore

#2 does NOT see a white hat on #3.
There can NOT be a white hat on #3 and so

There is a red hat on #3.

Number three figured this out
and correctly stated that his
hat was red.

Crates of Fruit Answer

*If you tell me to reach into the "Apples" box and
I pull out an Orange, you won't know if the box is full of
Oranges or if the box is full of Apples and Oranges.
The same thing is true if you tell me to reach into
the "Oranges" box and I pull out an Apple.*

*Therefore, you must tell me to reach into
the "Apples and Oranges" box.
Whatever fruit I pull out, that's the fruit that is in that box.
Let's say I hand you an Apple.
You would remove the "Apples and Oranges" label
and put it on the table. Take the "Apples" label and put it
on the box with Apples. Put the "Oranges" label on the box
that used to say "Apples" and put the "Apples and Oranges"
label on the box that use to say "Oranges".*

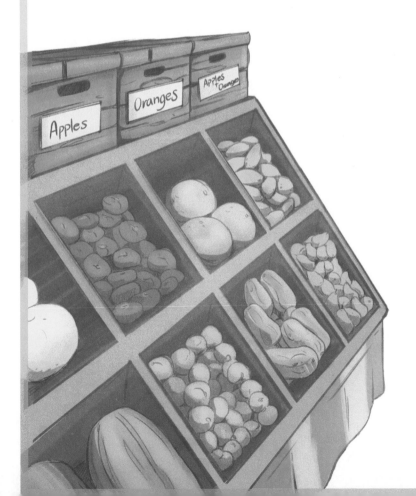

A Ship and Its Boiler Answer

This is a math riddle. To solve it we need to use simultaneous equations. To begin, find two equations.
It helps tremendously if you first build a small chart.

	Ship	Boiler
Now	2x	y
Then	y	x

The riddle refers to the ages of the ship and the boiler over time.
So the chart refers to their age Now and in the past, Then. Let's call the past age of the boiler "x". Then the ship's current age is 2x
("A ship is twice as old as its boiler was").
Let's call the present age of the boiler "y".
Then the ship's past age is y ("when the ship was as old as the boiler is").

Now we can easily find two equations.
$2x + y = 49$ is obvious.
The second equation comes from realizing that the difference in age between the ship and the boiler is constant, now vs. then.
In other words if five years have gone by, between Then and Now, then $2x - y = 5$ and $y - x = 5$. As an equation it looks like this
$2x - y = y - x$.
Let's rewrite that equation.

$2x - y = y - x$

$2x - y - y + x = 0$

$3x - 2y = 0$

Now take both equations and solve simultaneously.
$2x + y = 49$

$\underline{3x - 2y = 0}$

A Ship and Its Boiler Answer (cont.)

We're going to add the lower equation to the top equation.
We need to get rid of either the x or the y.
If we multiply the top equation by 2, y will become 2y.
Then when we add 2y to -2y, we'll get rid of y.

$2[2x + y = 49]$
$\underline{3x - 2y = 0}$

$4x + 2y = 98$

$\underline{3x - 2y = 0}$

$7x = 98$

$x = 98/7$

$x = 14$ *(Per the chart, this is the past age of the boiler.)*

The ship's current age is 2x = 28. The current sum of their ages is 49.
The current age of the boiler is 49 - 28 = 21.

The ship is 28 and the boiler is 21.

Four Pieces of Cardboard Answer

#1 If number 1 has a circle, then is proves that every red one does not have a square.
  Number 1 is needed.

#2 If number 2 has a square or a circle, it has nothing to do with red.
  Number 2 is not needed.

#3 If number 3 is red, then it proves that every red one does not have a square.
  Number 3 is needed.

#4 If number 4 is green, it tells us nothing about red. If number 4 is red, it does not prove that every red one does not have a square. Most people incorrectly include number 4 in their answer.
  Number 4 is not needed.

So the answer is:
We must pick up #1 and #3, in order to answer the question,
"Does every red one have a square on its other side?"

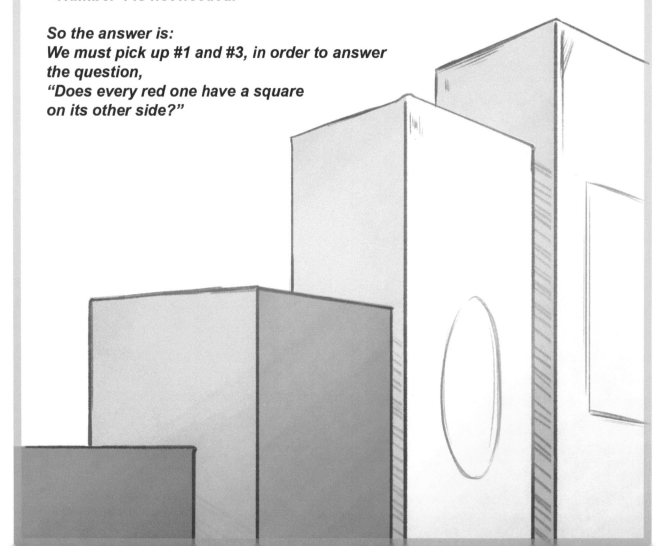

Diamonds and Rubies Answer

The Princess reached into the box and pulled out a ruby.
She instantly dropped it into the garden that was littered with
diamonds and rubies and said,
"Oh no, I dropped it. That's O.K., we'll look in the box and
see what's left. If it's a diamond, then I had the ruby and
if it's a ruby, then I had the diamond."

The beautiful Princess married the handsome Prince
and they lived happily ever after.

Monkey Business Answer

Labels
BL=Banana's length in inches
BW=Banana's weight, ounces
RL=Rope's length, feet
RW=Rope's weight, pounds
KW=Monkey's weight, ounces
KA=Monkey's age, years
MA=Mom's age, years
WW=Weight of weight, ounces

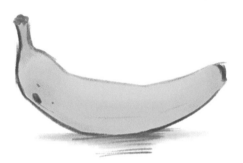

Equations
"The banana weights two ounces per inch."
1. BW (ounces) = 2BL (inches)

Solve for BL
2. BL (inches) = BW (ounces) / 2 (ounces)

******* IF WE KNOW WHAT THE BANANA'S WEIGHT (BW) IS,
WE CAN SOLVE THE RIDDLE***

"The rope is as long (in feet) as the age of the monkey (in years)."
3. RL (feet) = KA (years)

The rope "weighs one third of a pound per foot."
4. RW (pounds) = RL (feet) / 3 (pounds)

Rewrite, replace RL with KA, per # 3:
5. RW (pounds) = KA (years) / 3 (pounds)

The weight is "equal to the weight of the monkey."
6. WW (ounces) = KW (ounces)

*"The weight of the monkey (in ounces) is the same as
the age of the monkey's mother."*
7. KW (ounces) = MA (years)

Rewrite, replace KW with WW, per #6:
8. WW (ounces) = MA (years)

"The combined ages of the monkey and its mother are thirty years."
9. KA (years) + MA (years) = 30 (years)

Monkey Business Answer (cont.)

Solve for MA

10. MA (years) = 30 - KA (years)

"One half the weight of the monkey, plus the weight of the banana, is one fourth as much as the weight of the weight and the weight of the rope."

11. KW (ounces) / 2 + BW (ounces) = 1/4(WW (ounces) + RW (pounds))

 ***** IF WE KNOW WHAT THE BANANA'S WEIGHT (BW) IS,
WE CAN SOLVE THE RIDDLE *****

Solve for BW
12. BW (ounces) = 1/4(WW (ounces) + RW (pounds)) - KW (ounces) / 2

Everything is in ounces, except RW. Convert RW to ounces by multiplying it by sixteen. Replace RW (pounds) with 16RW (ounces)
13. BW (ounces) = 1/4(WW (ounces) + 16RW (ounces)) - KW (ounces) / 2
     Rewrite, replace WW with MA, per #8

     Rewrite, replace RW with KA/3, per #5

     Rewrite, replace KW with MA, per #7:

14. BW = 1/4(MA + 16KA/3) - MA/2

"The monkey's mother is half as old as the monkey will be when it is three times as old as its mother was when she was half as old as the monkey will be when it is as old as its mother will be when she is four times as old as the monkey was when it was twice as old as its mother was when she was one third as old as the monkey was when it was as old as its mother was when she was three times as old as the monkey was when it was one fourth as old as it is now."
15. MA = 1/2 * 3 * 1/2 * 4 * 2 * 1/3 * 3 * 1/4 KA

     MA = 3KA/2

Rewrite, replace MA with 30 - KA, per #10:

16. 30 - KA = 3KA/2

Monkey Business Answer (cont.)

Solve for KA

17. $2[30 - KA = 3KA/2]$

    $60 - 2KA = 3KA$

     $60 = 5KA$

      $12 = KA$

Per #10, we know that MA = 30 - KA, so plug in the value for KA

18. $MA = 30 - 12$

   $MA = 18$

Plug these values into #14; BW = 1/4(MA + 16KA/3) - MA/2:

19. $BW = 1/4(MA + 16KA/3) - MA/2$

   $BW = 1/4(18 + 16(12/3)) - 18/2$

  $= 1/4(18 + 16(4)) - 9$

  $= 1/4(18 + 64) - 9$

  $= 1/4(82) - 9$

  $= 20.5 - 9$

  $= 11.5$

$BW = 11.5$ ounces

Plug this value into #2; BL (inches) = BW (ounces) / 2 (ounces)

20. $BL = BW / 2$

    $= 11.5/2$

    $= 5.75$ inches

**** The banana is 5.75 inches long. ****

Misdirection Riddle Answer

I call this a misdirection riddle. The question is purposely misleading. The answer is finding how you are being misdirected. Concentrate on the amount of money in the register and you can't be led astray.

The men paid $10 each, so there is $30 in the register, for this transaction. The manager took $5 out, so now there is $25 in the register. We can track the original $30; $25 in the register and $5 with the bellhop. On the way to the men's hotel room, the bellhop puts $2 in his wallet. Now we have $25 in the register, $2 in the bellhop's wallet plus $3 in the bellhop's hand. The bellhop tells the men that the bill should have been $27 and he gives the men $1 each.

If the question read like the following, you wouldn't be fooled, "Each man originally paid $10 and each man received $1; therefore each man paid $9. $9 x 3 equals $27. Where is the missing $2.00?" You would instantly say, "In the bellhop's wallet." Instead I finished, "$9 x 3 equals $27 plus the $2 in the bellhop's wallet equals $29. What happened to the missing dollar? I mislead you. If I was honest and said, "$9 x 3 equals $27 minus the $2 in the bellhop's wallet equals $25," there would have been no riddle.

CPSIA information can be obtained
at www.ICGtesting.com
Printed in the USA
LVHW071251220620
658693LV00005B/8

* 9 7 8 1 0 8 7 8 5 4 2 8 1 *